Do we need Nuclear Power in the UK?

Riyaad Moreea

Cranmore Publications

A catalogue record for this book is available from the British Library

ISBN: 978-1-907962-35-6

Published by Cranmore Publications

Reading, England

www.cranmorepublications.co.uk

This book is dedicated to my amazing mum who has always been there for me. I would also like to thank my dad, sisters, step-mum, cousins and friends who are all an important part of my life.

Contents

Preface

As the finishing touches are being made to this book a potential nuclear disaster could be about to occur in Japan. There was immense damage caused to nuclear power plants located on the east coast of the country resulting from the earthquake and tsunami which affected the country on Friday 11 March 2011.

This event makes a consideration of our use of nuclear power a timely one. In this book I consider the advantages and disadvantages of nuclear power and whether we need to use this energy source in the UK.

Introduction

An ever increasing population has increased the amount of energy used in the UK throughout the last century and this trend is set to continue into the future. Energy produced from fossil fuels does not fit into the Government's carbon reduction targets. This means that an alternative cleaner source is required to meet the UKs future energy needs.

The UK is trying to move towards an energy supply that is produced from renewable energies, encouraged by government subsidies, but the potential for the different renewables depend on the UKs climate. Large scale wind farms and biomass plants look set to provide the most renewable energy in the UK along with energy contributions from small scale solar, hydro and geothermal sources.

Nuclear plants are also set to continue to provide energy in the UK as they provide a non variable supply of cheap electricity.

I will be contending that a combination of both nuclear and renewable energy is required to provide a stable and clean supply of energy in the UK, and that renewables should provide a greater percentage of electricity as load factors and installed capacities increase.

Chapter 1

Nuclear Energy in the UK

Nuclear energy currently provides about 14.2%, 2608 billion kWh (1), of the worlds electricity but as the worlds energy demand is predicted to increase by about 60% by 2030 (2), should nuclear power be used in the UK to meet this demand?

Some people may think that the UK should simply keep its energy use down, but this is an almost impossible task as energy is important to support economic growth, as well as the fact that the population will have increased from 1.1 billion in 1851 to a predicted 9 billion by 2050 (3).

The worlds energy consumption is expected to increase the most in developing countries and it is important to avoid and substitute the use of fossil fuels to meet these increases as growth based on fossil fuels is not assured to be feasible, due to the current fossil fuel price and this would also lead to a large increase in greenhouse gas emissions.

This book aims to look into the use of nuclear power not only in the UK but also worldwide to see if nuclear power is the best option for the additional energy generation capacity required for the UK. The book will also be looking into different renewable energies to allow comparisons to be made, giving a better understanding of the alternative ways to meet this demand.

Subsidies play an important role in the financial viability and therefore the construction of energy plants, so these will be discussed to show how the government are encouraging the building and use of new energy sources.

Construction times, load factors and costs of electricity generation are important factors when comparing different energy supplies, so these will be discussed and analysed to provide clear and comparable data to allow conclusions to be made on the future use of energy in the UK.

Another important issue is the tough targets set by the government of reducing carbon dioxide emission in the UK by 34% by 2020 and 80% by 2050. Different sources of electricity production have different carbon (and other greenhouse gas) emissions so these will be discussed to see how they fit in with reaching these targets.

Over the past 10 years renewable energy production has increased by over 3 fold but can it play an important role in meeting the UKs future needs? The main renewable energies currently used are wind, hydro, solar and biomass and these will be discussed to see whether theses renewables have the potential to meet the UKs increasing energy needs.

Chapter 2

The World Energy Scene

Pre-industrial energy demand was provided mainly by man and animal power and partially from the burning of wood for heating and cooking. The discovery of resources of coal along with technological advances fuelled the industrial revolution, including steam engines and improved transportation; there was also a rise of oil exploration and use. Oil and coal played a huge part in the Second World War and post-war industrial expansion such as the large growth in private cars. In more recent times natural gas, coal and oil have been used to a great extent for economic growth, as it is used to generate electricity which is now widely available. There has been a worldwide uptake of electricity, due to its flexibility and ease of distribution, and demand is continuously growing driven by consumer electronics, industrial activity and the increasing access in developing countries. Energy can be considered in two categories:

Primary – This is energy in the form of natural resources such as coal, oil, natural gas, wind, sunlight. Primary energy can then be split into either renewable energy sources such as solar, wave, wind or non renewable energy sources such as coal, oil, gas, uranium.

Secondary – This is energy in more usable forms which primary energy can be converted into such as electricity and petrol.

The world energy scene has changed greatly in the past few decades, especially after the oil price shocks in 1973 and the early 1980s. The popularity of coal as an energy source, until the mid 1960s, was sharply reduced due to the discovery of oil which had a lower price, greater cleanliness and greater ease of trade. During the past 2 decades there has also been an increasing output of nuclear and hydro energy as they can conveniently be used for electricity generation.

The total world primary energy consumption was estimated at around 469EJ in 2009 (1) and with a population of 6.8 billion this makes the annual average energy consumption per person about 69GJ. The world's energy consumption has been increasing since 1982 until a recent decline of 1.1% in 2009, mainly due to the recession. The latest figures showed that OECD energy consumption declined by 4.7%, North America declined by 4.5%, Europe declined by 5% and CIS declined by 8.5% in contrast to less developed countries such as Asia, which showed an increase of 4%. China is the world's largest energy consumer (18%), an increase of 8% during the year.

The majority of energy comes from the combustion of fossil fuels. These fuels are used for: Electricity (38%), Industry (19%), transportation (19%), and residential (24%).

Demand for primary energy has grown for coal, oil, gas, hydro, and nuclear, but at different rates. Overall worldwide energy consumption has increased by 60% over the last 25 years, with oil accounting for the highest amount (34.8%), followed by coal (28.2%), natural gas (25.0%), hydroelectricity(8.0%) and nuclear energy (4.0%). A very interesting point to note is that over the past 25 years the energy consumption per person has declined from 69.9GJ in 1984 to the current value of 69GJ.

Electricity generation worldwide has increased by 380% over the last 36 years. When we look at the sources it is clear that coal/peat provides the most electricity. A combination of renewables and waste started to make up a small percentage around 1990 and has slowly increased. Geothermal, solar and wind started to make up a small percentage of the worlds electricity generation in 2004 and produces roughly the same percentage as the combination of renewables and waste in 2008 after only 4 years.

North America and the Far East have the highest total energy consumption. The Far East uses more solid fuel to produce their energy than the total energy produced by all the fuel types in Latin America, Eastern Europe, Africa, Middle East & South Asia, South East Asia & the Pacific. North America also produces the most energy from nuclear power, while Western Europe produces the most energy from renewables.

Meeting the worlds growing demand for energy requires investment in infrastructure, as well as continuous technological advancements which allow access to extra resources. Energy producers will play an important role in meeting future energy challenges as they decide how they produce their energy; consumers will also play an important role as they decide how they use the energy.

Chapter 3

Use of Nuclear Power Worldwide

Nuclear technology uses energy released when atoms of certain elements are split. This technology was first developed in the 1940s, with research initially focussing on producing bombs by splitting uranium or plutonium atoms during the Second World War.

Generating nuclear power for electricity started more than 50 years ago and now produces as much global electricity as was produced at that time by all sources. Today there are nearly 440 nuclear reactors producing 376 GW (14%) of the world's electricity, down from a figure of 18% 10 years ago, with more than 15 countries relying on nuclear power for 25% or more of their electricity. Nuclear power accounts for approximately 6% of all primary energy production. Four countries produce more than 50% of their electricity using nuclear power: Belgium (51.7%), France (75.2%), Lithuania (76.2%), and Slovakia (53.5%).

Nuclear electricity production has increased from 50TWh in 1971 to a peak of 2600TWh in 2006 and has slightly declined to the current level of 2500TWh. The reason for this decline was due to the closure of power plants in Germany, UK and Japan. The increase over the past 40 years is not only due to an increase in nuclear power plants but also to an increase in performance from existing power plants. For example there was an increase of 210 TWh of electricity production between 2000 and 2006 but there was no net increase in the number of reactors. It is estimated that one quarter of the worlds reactors have a load factor of more than 90% and almost two thirds have a load factor of 75% or above compared with a quarter of them in 1990.

Large increases in nuclear power plants between 1970 and 1990, especially in developed countries, were followed by many cancellations of projects leading to a slowdown in nuclear energy development. This slowdown was due to the electricity market deregulation, slower growth of electricity demand and negative public perception due to accidents such as Three Mile Island in 1979 and Chernobyl in 1986.

It is estimated that there are 127 nuclear plants that are over 30 years old (2) and about 75% of the worlds plants will have to be replaced by 2030 to maintain the power supply from the nuclear sector.

This would mean 14 new reactors would have to be built every year, but this seems unlikely to be feasible as there are 58 reactors under construction (3) with scheduled starts of operation in 5-7 years, which amounts to 8 new reactors a year.

Recently there has been an increase in interest and development of nuclear energy due to rising gas prices, energy security concerns and because it is seen as a clean alternative to fossil fuel power plants. Approximately 58 reactors are being constructed in 15 countries mainly China, South Korea and Russia. It is estimated that there will be at least a 73GW net new capacity by 2010 and 511GW net new capacity by 2030(4).

In the UK there are 19 operating nuclear reactors at 10 power stations providing 17.9% of the UKs electricity. Only three of these stations are scheduled to be in operation by 2020 equating to a 7.4GW reduction of generating capacity. UK ministers believe that non fossil fuel energy production is required to meet the countries energy requirements to ensure that Britain is not over-dependent on foreign sources of energy, such as the Middle East or Russia, as North Sea oil and gas runs out. Cleaner energies are also seen as ways of helping Britain to meet its carbon reduction targets and help fight climate change.

Advantages of Nuclear power

- Nuclear power plants do not require a lot of space, compared to a lot of alternative energy production sources. This allows them to be built in developed areas of land meaning that the power does not have to be transferred over long distances, minimising energy losses.

• It is a clean form of energy production in relation to greenhouse gas emissions, as it does not depend on the burning of fossil fuels.

• Nuclear energy is an extremely concentrated form of energy so large quantities can be produced in short periods of time.

• Nuclear reactors are more reliable than most other energy plants, as they don't rely on fuel/weather conditions in the same way and are not as vulnerable to fluctuations in the market.

• The uranium needed for nuclear power is found in large quantity in both earth and seawater and is available at a viable cost.

Disadvantages of Nuclear power

• Nuclear power generates radiation, which can harm people.

• Nuclear meltdowns can occur which release large amounts of radiation into the surrounding area causing severe devastation (e.g. Chernobyl 1986).

- Radioactive nuclear waste is produced as a waste material, which cannot simply be disposed in the ordinary way. The fuel is toxic for very long periods of time and handling/disposal is an ongoing environmental issue. The waste is usually stored in the plants, kept underground or shipped to different countries

- Some reactors produce plutonium, which could be used to produce nuclear weapons, so there is a danger of terrorists stealing the plutonium to make nuclear weapons

Chapter 4

UK Government Subsidies for Nuclear Power and Renewables

Over the past 15 years there have been subsidies for nuclear power in the UK that have taken the form of exemptions from costs that a business is usually required to cover, and these subsidies usually come from the taxpayer. These exemptions include:

1. Limitations on liabilities – Nuclear power plants only pay a small percentage of the cost of insuring against disasters such as Chernobyl. The 1982 Energy Act legislates

that nuclear power plants must insure up to a liability limit of £140 million, although amending protocols are looking to increase the liability limit to €1500 million (5). This subsidy of reducing the true cost of insuring against a nuclear disaster makes nuclear plants more commercially viable.

2. Underwriting of commercial risks – The nuclear industry, similar to the banking industry, have their profits privatized but the risks are socialized. The UK government underwrites most of the commercial risks of nuclear plants, so if they get into financial difficulty the government will bail them out. For example, in 2005 the UK government

bailed out British Energy at a cost of approximately £5 billion (6).

3.   Short to medium cost of disposing of nuclear waste – The nuclear industry is paying a much lower short to medium cost to dispose of nuclear waste. The government is helping to meet the expensive decommissioning and waste disposal costs by providing a fixed price contract. This reduces the full commercial price which would make disposal extremely expensive.

4.   Support for the nuclear industry – There are various offices, staff and institutions that

support the nuclear industry which are funded using public funds. The Nuclear Decommissioning Authority (NDA) uses about 50% of the Department of Energy and Climate Change's £3 billion budget, where 58% of the NDAs £2.78 billion budget comes from public funds (7). Other UK facilities for the nuclear industry funded by public funds include(8): National Nuclear Laboratory (NNL), Office for Nuclear Development (OND), Nuclear Advanced Manufacturing Research Centre (NAMRC).

In October 2010 Minister of State for the *Department of Energy and Climate Change,* Charles Hendry, stated support for new nuclear capacity but without subsidies. He also stated "The coalition agreement clearly sees a role for new nuclear, provided that there is no subsidy. We are clear. It is for private sector energy companies to construct, operate and decommission new nuclear plants." This suggests that subsidies for new nuclear plants in the UK will be greatly reduced.

The UK government provides support for renewable energies through feed-in tariffs, renewable energy credits, tax credits, cash grants, research and development and other direct subsidies. In 2010 subsidies for renewable energies had risen from £7 in 2007 to £13.50 per year on average household electricity bills, which totalled £1.04 billion during 2008-2009 (9).

Under the Energy Act 2008 extra incentives are to be offered to encourage the use of offshore wind, biomass and other renewable technologies, it is estimated that £6 billion will be available by 2020 to provide financial incentives (9).

Since April 2010 the UK government has revealed a feed in tariff scheme to incentivise low carbon technologies such as photovoltaic panels and wind turbines (Up to 5MW), where the owners are paid for the electricity they produce even if this is used up by their own house. This scheme guarantees a minimum payout for all electricity generated as well as a separate payment for the electricity exported to the national grid.

Renewable obligation certificates place an obligation on UK suppliers of electricity to source an increasing proportion of their electricity from renewables. These are designed to incentivize renewable generation as suppliers that do not have enough certificates must pay a fine.

Chapter 5

Electricity Supply and Greenhouse Gas Emissions

In 2009 2608 TWh of electricity was generated by nuclear energy worldwide and 62.9 TWh was generated in the UK.

The calorific value of coal, oil and gas are 26.1GJ/tonne, 45.7GJ/tonne and 39.6MJ/cubic meter respectively (10). All data and calculations can be found in the Appendices.

Table 1 - Comparison of electricity produced by coal, oil and gas

	Coal	Oil	Gas
Calorific Value(GJ/ton)	26.1	45.7	39.6
Thermal energy content (kWh/ton or cubic meter)	7250	12694	11000
Thermal efficiency of energy conversion (%)	35	42	51
Electricity generated (kWh/ton or cubic meter)	2537.5	5331.48	5610
Amount required to replace worldwide nuclear electricity supply (Giga ton or cubic meter)	1027.78	489.17	464.89

Amount required to replace UK nuclear electricity supply (Giga ton)	24.79	11.80	11.21
UK Sulfur Dioxide (tons) - Main cause of acid rain	74370	35400	60480
UK Nitrogen Oxides (tons) - Causes smog and acid rain	29748	14160	8624
UK Carbon Dioxide (tons) - Greenhouse gas suspected of causing global warming	9420200	4484000	3472000

Table 1 shows that oil is the most energy intensive fossil fuel followed by gas, then coal. Gas fired plants are the most efficient even though 51% does not seem very high – this is due to a lot of inefficiencies during the energy conversion process. One ton of oil produces 11% more electricity than the same quantity of coal. If the UKs nuclear electricity generation was produced using fossil fuels then coal power plants would produce the most carbon dioxide (greenhouse gas that causes global warming), sulphur dioxide (main cause of acid rain) and nitrogen oxides (causes smog and acid rain) emissions – a total of 110% more than oil and 166% more than gas.

When carbon emissions are considered the lifecycle emissions of the energy sources should be considered as these include emissions arising from factors such as, mining, transport, processing, construction, operation, maintenance, and decommisioning. The highest carbon footprint of nuclear energy produces 5 gco_2/kWh, which is the same as wind energy. This figure is a lot lower than the other low carbon energy sources, almost 75 gco_2/kWh less than biomass, 545 gco_2/kWh less than photovoltaics, 455 gco_2/kWh less than marine and 55 gco_2/kWh less than hydro.

The UK government have introduced a number of measures to reduce greenhouse gas emissions to help reach their targets of a 34% reduction by 2020 and a 80% reduction by 2050.

The UK government uses an Emissions Trading Scheme which gives industries an allocation of how much carbon dioxide they can emit to try and reduce greenhouse gas emissions. If a company emits more than they are allowed then they must purchase permits to make up for the extra and if they emit less they can sell their unused allocations.

The UK Renewables Obligation forces retailers to buy a certain percentage of the electricity (9.1% in 2008) they supply from renewable sources otherwise they will be charged a penalty.

A Climate Change Levy has been introduced to help increase energy efficiency and reduce emissions of greenhouse gases. Companies pay the levy (0.43 p/kWh) via their energy bills for electricity, natural gas, petroleum, coal and coke.

The Carbon Emissions Reduction Target requires every domestic energy supplier with more than 50000 customers to deliver measures that will provide overall lifetime carbon dioxide savings of 293Mt by December 2012. The suppliers are trying to meet these targets by promoting the uptake of low carbon energy solutions including supplying free energy efficient light bulbs and providing grants for insulation. It is estimated that the cost to suppliers of achieving this target, from April 2008 to December 2012, will total £5.5 billion (11).

A couple of other plans for electricity suppliers to reduce their greenhouse gas emissions include:

Carbon dioxide capture and storage – This process separates the carbon dioxide from energy related sources and stores it underground.

Carbon offset – This is usually achieved through financial support of projects that decrease greenhouse gasses e.g. forestry projects, destruction of landfill methane, destruction of industrial pollutants.

Chapter 6

Construction Times, Load Factors and Costs of Electricity Generation

Load factors are a measure of an output compared to the theoretical maximum output that can be produced over a period of time. A higher load factor is better as this usually means a higher output and a lower cost per unit, as the total fixed costs of a plant are distributed over a larger output.

The load factor is typically worked out on a monthly or annual basis as it continuously changes. They are helpful for estimating quantities of electricity at a power station, so you can predict how many units of electricity will be supplied and this can be used to balance supply and demand.

Figure 1: Estimated load factors of energy technologies (21)

Energy Technology	Load factor
Sewage Gas	90%
Farmyard Waste	90%
Energy Crops	85%
Landfill Gas	70-90%
Combined Cycle Gas Turbine (CCGT)	70-85%
Waste Combustion	60-90%
Coal	65-85%
Nuclear Power	65-85%
Hydro	30-50%
Wind Energy	25-40%
Wave Power	25%

Figure 1 shows the estimated load factors of various energy technologies and it is clear that hydro, wind and wave power have the lowest values although it varies according to location/technology and also it must be noted that the load factor is the output compared to the theoretical maximum, not its average. This means that a wind farm with a load factor of 25% and an installed capacity of 10MWh will produce 2.5MWh, i.e. 25% of what it would produce if it was operating at maximum output. The load factor is very important when looking into building energy technologies as the manufactures capacity may be different to the actual capacity, thereby altering energy calculations in a way which might not make a project viable.

Different types of power generation technologies take varying periods of time to construct due to factors such as: size of plant, quantity of materials required, difficulty with locations, man power required, and imported items. Energy projects that have long construction times are normally capital intensive, but usually produce the cheapest energy, otherwise the economics of the project would not be justified.

Figure 2: Construction times of different power generation technologies (22)

Power generation technology	Construction time (months)
Nuclear pressurized water reactor	36-60
Nuclear boiling water reactor	36
Combined cycle gas turbine	36-40
Coal plant	48-60
Oil fired power plant	24-36
Biomass fired plant	24-30
Wind –onshore	16-20
Wind offshore	18-30
Hydroelectricity	60-120
Solar	12-36

Figure 2 shows that nuclear power plants generally take longer to construct than renewable energy power generation systems. Solar and wind systems are the fastest to construct while hydroelectricity systems can potentially take the longest time – up to 10 years.

Figure 3: Plant life of power generation technologies (23)

Power generation technology	Plant life (years)
Coal plant	40
Nuclear	40
Oil fired power plant	25
Gas fired power plant	25
Solar	25
Wind	25
Biomass	35

Figure 3 shows that nuclear and coal power stations have the longest plant life of 40 years followed by biomass plants (35 years); oil fired plants, gas fired plants, solar and wind all have a plant life of 25 years. The plant life is also an important factor when planning the construction of a power plant as it will play a huge part in the total amount of energy produced throughout its lifetime and therefore the financial viability. It allows the timing of actions to be planned, so when a plant reaches the end of its lifecycle decisions can be made e.g. deconstructing the plant, extending the lifetime with extra work.

One of the most important factors when planning the construction of an energy plant is the cost, as the developer wants to make a profit from the project. All of the factors above have an influential effect on the finance of a project.

Figure 4: Financial costs of power generation technologies

	Fuel Costs (£ per GJ), unless stated	Operation and maintenance (% of construction cost)	External costs (% of construction cost)	Construction costs per Kilowatt (£/KW)	Decommisioning	Cost per KWh of electricity produced (p)	Reference
Nuclear	0.18	3	0.016	1150 - 2800	100 million – 400 million	2.3	(22) (24) (25) (26)
Coal power plants	1.39	3	0.20-0.43	800-1000	-	2.6	(22) (24) (26)
Oil fired power plants	3.29	2	0.2-0.3	600-1000	-	6.7	(22) (24) (709)
Gas fired power plant	2.05	3	0.06-0.12	330- 700	-	3.1	(22) (24) (26)
Wind farms	0	Onshore –3 Offshore - 6	.009	Onshore 600-750 Offshore 800-1000	minimal	On-shore – 3.7 Off-shore – 5.5	(706) (24) (26)
Bio-mass	Poultry litter - £7 per ton Short-rotation coppice (oven-dried) - £40 per tone	5	.062	750 - 1700	minimal	2.32-6	(706) (24) (26)

	Wood pellets - £58 - £73 per ton						
Solar	0	0.15	0.018	3500	minim-al	30	(23) (24) (26)
Hydro	0	3	0.037-0.1	1000-2500	-	3.5	(23) (24) (26)

Figure 4 shows that the fuel costs of renewable energies are free except for biomass, where the fuel cost ranges from £7-£73 depending on the type. Nuclear fuel costs are quite low due to the low cost of uranium and the most expensive fuel type is for oil plants, followed by gas and coal plants.

Solar power has extremely low operation and maintenance costs due to no moving parts and the cells requiring little upkeep. The operation and maintenance costs of the other technologies range from 2%- 6% of the construction cost. Offshore wind farms have twice the operation and maintenance costs compared to onshore wind farms due to exposure, accessibility, and the specialist equipment required, e.g. boat cranes.

Coal and oil have the largest external costs and wind has the lowest. Solar and nuclear have very similar costs, nuclear is slightly lower, and both are lower than gas and biomass. Hydro electricity has the largest range, which reflects the very site specific nature of hydropower.

The UK domestic electricity price including tax is 9.1665 p/kWh (12) and the industrial electricity price tax is 7.7 p KWh (12). The external cost of wind power is estimated to be about 30 times lower than the industrial cost of electricity (13), which makes the external cost of wind a small fraction of total electricity costs. At the other extreme, the external costs of coal fired electricity are higher than the electricity price charged to the industrial sector (13); so the external cost of coal generation is a dominant factor in the price of coal generated electricity. The carbon tax, which is a large part of the external cost of coal and gas, is used by the government to deter the burning of fossil fuels as mentioned in *Chapter Five*.

It is clear that even though nuclear has the highest starting up costs, the low price of uranium makes it able to compete with the cost of electricity produced.

Chapter 7

Growth in Renewables over the Past 10 Years

The UK produces 6.1% (14) of its overall energy and 8% (14) of its electricity from renewables; this is up from 2.6% in 2000. Although this seems like it makes up a small percentage of the UKs electricity use the amount of electricity generated from renewables has increased by 321% over the last 10 years.

Wind and wave power has had the largest increase in tonnes of oil equivalent with a 99900% increase over the past 10 years. Land fill gas and other biomass have had a similar increase of 1952% and 1890% respectively. Geothermal and active solar heating has had a 900% increase but Hydro power has only had a 1% increase over the same period.

One of the main reasons that the UKs electricity generation from renewable energy has increased is because the government has set itself a target of producing 20% of electricity from renewables by 2020.

The UK government has also set a target of reducing carbon dioxide emissions by 34% and 80% by 2050. These targets have led to a large increase in installation of renewables encouraged by government subsidies as mentioned in *Chapter 4.*

Let us consider the breakdown of current UK renewable energy sources. Biomass is the most used renewable energy source with 80.7%, followed by wind with 11.6%, large scale hydro with 5.8%, and geothermal, active solar heating and small scale hydro making up the remaining 1.8%. Out of the 6.875 Ktoe, 4.9 Ktoe was used for electricity generation, 1.01 Ktoe was used for road transport and 0.97Ktoe was used to generate heat.

Installed PV systems connected to the grid in the UK have been growing steadily over the last decade and produced 14 MW in 2006. This is a relatively slow uptake compared to the other renewable energies and this is mainly due to the lack of sunlight in the UK, which affects the economics of installing a system as they have long payback periods.

In 2007 wind energy took over hydropower as the UKs second largest renewable generation source, making up 2.2% of the UKs electricity supply. The UK has some of the fastest onshore and offshore wind speeds in Europe, so it seems that the greatest potential of increasing the UKs renewable energy is from wind. The government is aware of this potential and is starting to exploit it as there are a lot of wind production sources under construction, consented to or in the planning stage.

The performance of biomass plants depends on the fuel being used, the technology used and the size of the plant. In the UK biomass electricity is mainly generated by landfill gas and refuse derived fuels, this not only reduces land fill waste but also does not require crops to be grown or imported. Unlike some other renewables biomass has a controllable supply irrespective of weather conditions, so it can supply both peak load and base load power.

Another type of non polluting technology that looks set to play a huge role in reducing carbon emissions in the future are fuel cells. Fuel cells have the potential to replace a lot of today's polluting energy converters, reducing pollution considerably as the only by products are heat and water. By combining electricity and heat output, fuel cells can achieve around 80% efficiency, which is a lot greater than all of the technologies mentioned above. There are technical, economic and market factors that have prevented the technology becoming widely available. These include the problem of obtaining hydrogen, a high cost making it not commercially viable, and a currently low durability.

Chapter 8

Discussion

Even though worldwide energy consumption reduced last year, world energy production has increased by 60% over the last 25 years, mainly fuelled by fossil fuels. It is clear that the predicted energy increase cannot be met by using fossil fuel plants, as the UK government has a tough carbon reduction target that can only be achieved using cleaner fuel sources. Nuclear power can be used to meet these carbon targets and energy needs, but another option that is available is renewable energy.

The UK government's target is to cut down carbon emissions in the UK by 34% by 2020 and by 80% by 2050. If coal plants were used to supply the UKs energy needs in 2009 then 9.5 Mega tons of SO_2, NO_x and CO_2 would have been emitted. Oil fired plants would have produced 4.5 Mega tons of SO_2, NO_x and CO_2 and gas fired plants would have produced 3.5 Megatons of the same greenhouse gases. It is clear that fossil fuel plants produce a lot of greenhouse gases when they are burnt to produce energy, so these cannot be used to meet the UKs increasing energy needs otherwise the government targets will not be met, whereas nuclear and renewable technologies do not emit these gases.

The UK currently has 19 nuclear reactors but there are no reactors being built at the moment, this is partly to do with the subsidies the government provides. The *Department of Energy and Climate Change* stated that there will be no subsidies for new nuclear power generation, compared to a lot of exemptions that the government gave to nuclear power plants in the past. This makes them less financially viable projects.

Electricity generated in the UK from renewable energy sources has increased from 6 TWh to 26 TWh in the last 20 years, with the largest increases coming from wind and biomass energy. This increase has been boosted by government subsidies, which have made renewable projects more viable and attractive to investors. The feed in tariffs have made small scale low carbon technologies more financially attractive which has increased the number of installations in the domestic sector. The renewable obligation certificates have increased the proportion of renewable energy used in the industrial sector, with the aim of encouraging energy suppliers to source a certain percentage of their energy from renewable sources.

Biomass plants have the highest load factor (70%-90%), nuclear power plants also have a high load factor of between 65%-85%. This means that they can supply electrical energy at a lower cost, as the number of units generated for a given demand increases; therefore the overall cost per unit generated reduces. Wind (25%-40%) and hydro energy technologies (30%-50%) have lower load factors, which leads to greater load variations and a lower energy production compared to the theoretical maximum. This means much larger plants, in terms of power output, are required to generate the same amount of electricity as nuclear plants.

Nuclear plants tend to take longer to construct then wind farms, biomass plants and solar farms, but not hydropower plants, mainly due to the scale of the projects. The construction time is an important factor, as it not only impacts on the financial viability of the project but also how long it will take before energy can be used from the technology. Any overruns in the construction time can delay power outputs causing expensive costs when the electricity grid is short of capacity or is supplying power from plants with high variable costs.

Although nuclear plants take the longest to construct they do last the longest (40 years), followed closely by biomass plants (35 years) and solar and wind farms which have an estimated life of 25 years. A longer plant life means that energy will be produced for a larger period of time, which increases the time in-between new plants having to be built as a replacement, which involves large start up costs.

Nuclear plants are expensive to build, in terms of costs per kilowatt, compared to renewable technologies, but they do provide the cheapest rate of electricity even though renewable energy fuel costs, except biomass, are free. The main reason for this is due to the low load factors of the renewable plants and variability in energy production. The decommissioning costs of renewable technologies are very low whereas nuclear power plants have extremely large decommissioning costs involving the safe management of nuclear materials, decontamination, plant dismantling, demolition and site remediation.

A small percentage of the UKs electricity use (8%) currently comes from renewable energies, with biomass, wind and hydropower producing over 96% (14). The UK has great potential for the development of wind farms as it has some of the fastest wind speeds in Europe. It is estimated that the UKs offshore resources that are technically available are 3 times the total amount of the UKs electricity consumption (15). Biomass could also play a large part in the future of the UKs power as it generates a constant supply of energy in all weather conditions from crops, trees and waste. There is a limited potential for large increases from hydro power as most of the current production comes from large dam projects installed many years ago, and there are not many large practical locations. However, there is still untapped small hydropower potential in certain parts of the UK. Solar energy does not look like it will play a huge role in the future of UK energy as it is expensive to set up and there is likely to be limited energy production due to a lack of sunshine.

Nuclear power plants have the lowest lifecycle carbon emission, producing an estimated $5gco_2/kWh$. Most of these emissions come from the mining of the uranium and decommissioning of the power plants, but it has been suggested that as higher grade uranium ore deposits become scarcer lower grade ores will be used (16). This would increase the carbon footprint of nuclear generation as more energy would be used to extract and refine the uranium to a suitable level for the reactor.

Renewable energies usually have a higher carbon footprint compared to nuclear power due to economies of scale and to renewable technology parts being transported from various locations around the world.

Chapter 9

Conclusion

The IEA is estimating that the worlds demand for energy will increase by 281.4 EJ by 2030, which is a lot of extra power that has to be produced. One of the options in the UK for meeting this increased energy use is nuclear power. But is this the best way?

It is clear that drastic action is required in order for the UK government to reach its targets of reducing carbon dioxide emission in the UK by 34% by 2020 and 80% by 2050. The IEAs prediction that energy demand will increase by 60% by 2030 driven by the reality of an increasing population means that more clean energy is required. Both nuclear and renewable technologies will be required to provide clean power sources to meet these energy demands.

The government is trying to meet increasing energy needs by using renewables, which is evident as subsidies for nuclear plants are decreasing and many incentives are being introduced to promote the production of renewable energies both in the domestic and industrial sectors. Strategies such as the feed in tariff are likely to increase the installation of small scale renewable technologies, and other strategies such as renewable obligation certificates are encouraging energy suppliers to source more of their energy supply from renewable technologies.

The load factors of wind and hydro technologies are very low, so there is a limit to the energy produced in relation to the plant size, but as materials and processes become more efficient the load factors should increase in time. Biomass and nuclear have much higher load factors leading to a reduction in the overall cost per unit generated as more energy is produced.

The unpredictability of energy produced from renewable sources along with the low load factors make the electricity produced more expensive than nuclear power production.

The construction times of energy plants are extremely variable due to the nature of the specific projects but nuclear power plants tend to take longer to construct than all renewables except hydropower.

Due to the UKs wind resources wind farms can be used to provide more than the current level (3%) of the UKs electricity. Since it only produces energy on windy days, it can be used as an alternative for peak loads e.g. during daytime when it is windy and there is a high energy demand, but it is not reliable enough for base load requirements. The government should continue to encourage the building of both on shore and off shore wind farms.

Biomass energy production should also be increased as large scale biomass plants could provide stable energy production that is not dependant on the weather. The main disadvantages of biomass are carbon emissions if they are burnt and the fact that fuel is not available all year round so you cannot continuously feed in the materials; corn, barley, and wheat are seasonal crops and trees are a slow growing resource

It does not look like there will be any new large scale applications of hydro and solar power due to lack of suitable locations and costs; however, smaller scale applications could provide a small contribution to the UKs energy needs.

Nuclear power plants do provide the UK with a reliable source of energy production that produces electricity at a cheap price. It is the only viable non greenhouse gas emitting energy source that can provide energy 24 hours a day. This along with the fact that the plants do not emit greenhouse gases makes nuclear power look like a good source to meet the UKs future energy needs. The main reason why it is may not be seen as a good way to meet future energy demands is because of the radioactive waste the plants produce and the possibility of a nuclear meltdown.

Renewable energies are seen as a very clean way to provide energy, as they do not emit greenhouse gases and the fuel required is naturally occurring and does not require mining, except for biomass. Not producing radioactive waste makes renewables seem a better option than nuclear power, but the low load factors and uncertainty in energy production make them unable to provide all of the UKs energy needs.

Taking everything into consideration the UK needs a mixture of both nuclear and renewable energy. Nuclear energy provides cheap electricity that is less variable on a short and long term basis, and which is required to sustain economic growth and an increasing population. The high variability and lower load factors make renewable energy production unpredictable; it is, therefore, unlikely that these sources will completely replace nuclear plants and provide 100% of the UKs energy. However, their contribution to the UKs energy production should continue and as load factors increase and energy capacity increases they should reduce the UKs dependency on nuclear energy. This will reduce the amount of radioactive waste produced.

Bibliography

(1) DECC, 2009. Energy, [online] available at:

http://www.decc.gov.uk/assets/decc/Statistics/publications/dukes/307-dukes-2010-ch1.pdf

(2) ISEA, 2009. Worldwide Energy, [online] available at:

www.isea.org/programmes/a2/index.html

(3) WNS, 2010. World Nuclear Power Reactors, [online] available at:

 http://www.world-nuclear.org/info/reactors.html

(4) CBO, 2008. Nuclear Power's Role in Generating Electricity, [online] available at:

http://www.cbo.gov/ftpdocs/91xx/doc9133/05-02-Nuclear.pdf

(5) EF, 2010. Nuclear Subsidies, [online] available at:

http://www.mng.org.uk/gh/private/nuclear_subsidies1.pdf

(6) GO, 2009. Nuclear Incentives, [online] available at:

http://www.guardian.co.uk/business/2006/mar/17/nuclear industry.politics.

(7) TO, 2009. Nuclear Decommissioning, [online] available at:

http://business.timesonline.co.uk/tol/business/industry_sectors/utilities/article6717198.ece

(8) EF, 2010. Nuclear Subsidies, [online] available at:

http://www.mng.org.uk/gh/private/nuclear_subsidies1.pdf

(9) WNA, 2010. Energy Subsidies and External Costs,
[online] available at:

http://www.world-nuclear.org/info/inf68.html

(10) ATN, 2006. Calorific Values of Fuels, [online] available
at:

http://www.appletonlemoors.co.uk/docs/calorific_values.PDF

(11) CIAB, 2005. Reducing Greenhouse Gas Emissions,
[online] available at:

http://www.iea.org/textbase/nppdf/free/2005/ciab.pdf

(12) EON, 2010. UK Power, [online] available at:

http://www.ukpower.co.uk/home_energy/compare/dual/F
DVFRVCU/14481-EON-SaveOnline-4/details

(13) LB, 2009. The External Costs of Electricity Generation,
[online] available at:

http://lightbucket.wordpress.com/2009/06/24/the-
external-costs-of-electricity-generation/

(14) DECC, 2009. Energy, [online] available at:

http://www.decc.gov.uk/assets/decc/Statistics/publications
/dukes/307-dukes-2010-ch1.pdf

(15) GP, 2001. Offshore Wind, [online] available at:

http://www.greenpeace.org.uk/files/pdfs/migrated/Multim
ediaFiles/Live/FullReport/3443.pdf

(16) TEC, 2007. The Lean Guide to Nuclear Energy, [online] available at:

http://www.theleaneconomyconnection.net/nuclear/index.html

(17) WNA, 2010. Nuclear Share Figures, [online] available at:

http://world-nuclear.org/info/nshare.html

(18) WNA, 2010. Nuclear Power in the World Today, [online] available at:

http://world-nuclear.org/info/inf01.html

(19) IAEA, 2008. Energy, Electricity and Nuclear Power Estimates for the Period up to 2030, [online] available at:

http://www-pub.iaea.org/MTCD/publications/PDF/RDS1-28_web.pdf

(20) BBC, 2009. Nuclear Stations, [online] available at:

http://news.bbc.co.uk/1/hi/uk_politics/8349715.stm

(21) BWEA, 2010. Wind Energy, [online] available at:

http://www.bwea.com/energy/rely.html

(22) David, S. Biewald, B., 2008. Nuclear Power Plant
Construction Costs, Prism Press, UK.

(23) BMU, 2009. Electricity from Renewable Energy Sources,
[online] available at:

http://www.erneuerbareenergien.de/files/pdfs/allgemein/
application/pdf/brochure_electricity_costs.pdf

(24) WP, 2010. Cost of Electricity by Source, [online]
available at:

http://en.wikipedia.org/wiki/Cost_of_electricity_by_source

(25) NRC, 2007. Decomissioning a Nuclear Power Plant, [online] available at:

http://www.nrc.gov/reading-rm/basic-ref/students/decommissioning.html

(26) LB, 2009. The External Costs of Electricity Generation, [online] available at:

http://lightbucket.wordpress.com/2009/06/24/the-external-costs-of-electricity-generation/

(27) NEA, 2005. Projected Costs of Generating Electricity, [online] available at:

http://www.iea.org/textbase/nppdf/free/2005/ElecCost.pdf

(28)RED, 2009. National Renewable Energy Action Plan for the United Kingdom, [online] available at:

http://www.decc.gov.uk/assets/decc/what%20we%20do/uk%20energy%20supply/energy%20mix/renewable%20energy/ored/25-nat-ren-energy-action-plan.pdf

(29) DECC, 2010. Renewable Energy Growth, [online] available at:

http://www.decc.gov.uk/en/content/cms/statistics/publica tions/dukes/dukes.aspx

(30) BERR, 2008. Growth Potential for Renewable Electricity Generation, [online] available at:

http://www.decc.gov.uk/assets/decc/Consultations/Renewable% 20Energy%20Strategy%20Consultation/Related%20documents/1 _20090501125345_e_@@_TheGrowthPotentialforonsiterenewab leelectricitygenerationinthenondomesticsectorinEnglandScotland andW.pdf

(31) Toke, D., 2010. The UK Offshore Wind Power Programme: A sea-change in UK energy policy?, Energy Policy

(32) SS, 2010. Should We Use Nuclear Energy to Tackle Climate Change, [online] available at:

http://stephenstretton.com/a/2/5-ShouldWeUseNuclear.pdf

Appendices

Coal calculations

The thermal energy content of coal is 7250kWh/ton, and typical coal plant efficiency is 35%

Electricity generated per ton of coal is 0.35 x 7250 = 2537.5kWh/ton

Coal burnt to supply worldwide nuclear electricity energy = $2608 \times 10^{12} \div 2537.5 = 1.028 \times 10^{12}$ tons

Coal burnt to supply UKs nuclear electricity energy = $62.9 \times 10^{12} \div 2537.5 = 2.479 \times 10^{10}$ tons

SO_2 emissions = 2.479×10^{10} x 0.0000030 = 74370

NO_x emissions = 2.479×10^{10} x 0.0000012 = 29748

CO_2 emissions = 2.479×10^{10} x 0.00038 = 9420200

Oil calculations

The thermal energy content of oil is 12694kWh/ton, and typical coal plant efficiency is 42%

Electricity generated per ton of oil is 0.42 x 12694 = 5331.48kWh/ton

Oil required to supply worldwide nuclear electricity energy = $2608 \times 10^{12} \div 5331.48 = 4.89 \times 10^{11}$ tons

Oil required to supply UKs nuclear electricity energy = $62.9 \times 10^{12} \div 5331.48 = 1.18 \times 10^{10}$ tons

SO_2 emissions = 1.18×10^{10} x 0.0000030 = 35400

NO_x emissions = 1.18×10^{10} x 0.0000012 = 14160

CO_2 emissions = 1.18×10^{10} x 0.00038 = 4484000

Gas calculations

The thermal energy content of gas is 11000kWh/ton, and typical coal plant efficiency is 51%

Electricity generated per ton of gas is 0.35 x 7250 = 5610kWh/ton

Gas required to supply worldwide nuclear electricity energy = 2608x10^{12} ÷ 5610 = 4.64x10^{11} tons

Gas required to supply UKs nuclear electricity energy = 62.9x10^{12} ÷ 5610 = 1.12x10^{10} tons

SO_2 emissions = 1.12x10^{10} x 0.0000054= 60480

NO_x emissions = 1.12x10^{10} x 0.00000077= 8624

CO_2 emissions = 1.12x10^{10} x 0.00031= 3472000

Other books by the author:

The potential for ground source heat pumps in domestic houses in the UK

(2011)

www.ingramcontent.com/pod-product-compliance
Lightning Source LLC
Chambersburg PA
CBHW050551280326
41933CB00011B/1802